W9-CLE-829

NATURAL DISASTER!

All About Hurricanes

Discovering Earth's Wildest Storms

Cody Crane

Children's Press®
An imprint of Scholastic Inc.

Library of Congress Cataloging-in-Publication Data
Names: Crane, Cody, author.
Title: All about hurricanes / Cody Crane.
Description: First edition. | New York : Children's Press, an imprint of Scholastic Inc., 2021. | Series: A true book: natural disaster! | Includes bibliographical references and index. | Audience: Ages 8–10. | Audience: Grades 4–6. | Summary: "This book shows readers the awesome power of hurricanes"—Provided by publisher.
Identifiers: LCCN 2021003962 (print) | LCCN 2021003963 (ebook) | ISBN 9781338769647 (library binding) | ISBN 9781338769654 (paperback) | ISBN 9781338769661 (ebook)
Subjects: LCSH: Hurricanes—Juvenile literature. | Natural disasters—Juvenile literature.
Classification: LCC QC944.2 .C73 2021 (print) | LCC QC944.2 (ebook) | DDC 551.55/2—dc23
LC record available at https://lccn.loc.gov/2021003962
LC ebook record available at https://lccn.loc.gov/2021003963

10 9 8 7 6 5 4 3 2 1 22 23 24 25 26

Printed in the U.S.A. 113
First edition, 2022

Series produced by Priyanka Lamichhane
Book design by Kathleen Petelinsek
Illustrations on pages 42–43 by Gary LaCoste

Front cover: Background: A hurricane approaches the United States; top: A house is torn in half during Hurricane Sandy; top right: Hurricane Irma brings high winds and heavy rains to Florida; bottom: A volunteer helps rescue people stuck in a car during Hurricane Harvey.

Back cover: Multiple hurricanes approach the Caribbean islands and the United States.

Find the Truth!

Everything you are about to read is true ***except*** for one of the sentences on this page.

Which one is **TRUE**?

T or F The wind speeds of a Category 4 hurricane can be up to 145 miles per hour (233 kph).

T or F Scientists fly into hurricanes to collect data.

What's in This Book?

Hurricane Irma's high winds batter the Florida coast in 2017.

Post-hurricane flooding in Texas

A satellite orbits Earth as a hurricane spins below.

Washed Away

On September 8, 1900, Jacob Brooks and his cousin Allen were on a **beach** in Galveston, Texas. There, they saw something shocking. A **towering** wall of water was moving toward the island where their town was located. It was being pushed on shore by a monster **hurricane**.

That evening, the hurricane made landfall. Fierce **winds** tore through the city. Then came the **storm surge**. The huge wave Jacob and Allen had spotted earlier reached land. It **swallowed** the town in 15 feet (4.6 meters) of water.

A house shown tipped on its side after the Great Storm hit Galveston, Texas, in 1900.

By the time the hurricane was over, more than 8,000 people had lost their lives. It was the **deadliest** natural disaster in U.S. history.

People living in Galveston knew a **storm** was coming. They just did not know how bad it would be. Scientists didn't know how to give advance **warnings** about weather like they do now. Today, scientists can tell us when hurricanes are coming. Read on to find out more about these **powerful** storms and how scientists study and predict them.

The word "hurricane" comes from the Taíno Indigenous word *hurakán*, which means "god of the storm."

In 2017, Hurricane Irma struck Florida. This state is hit by more hurricanes than any other in the United States.

CHAPTER 1

Giant Ocean Storms

Hurricanes are huge, rapidly spinning storms. They form in the Atlantic and eastern Pacific Oceans during hurricane season, which runs from June to November. After forming, they usually move west, carried by strong ocean winds. Hurricanes pose the greatest risk to islands in the Caribbean Sea, the Gulf and East Coasts of the United States, and the east and west coasts of Mexico. But it is possible for these powerful storms to affect areas inland.

Storm Stages

A hurricane starts as a tropical depression, which is made up of a group of thunderstorms swirling around each other. A tropical depression can reach wind speeds of 38 miles per hour (61 kilometers per hour). If winds go from 39 to 73 miles per hour (63 to 118 kph), the depression turns into a tropical storm. A tropical storm turns into a hurricane if winds reach at least 74 miles per hour (119 kph).

Timeline of Record-Breaking Hurricanes

An engraving illustrating the 1780 Caribbean hurricane

Hurricane Diane

How a Hurricane Forms

Tropical depressions, tropical storms, and hurricanes normally develop over water that is at least 80 degrees Fahrenheit (27 degrees Celsius). The water **evaporates**, changing from a liquid to a gas. The warm, moist air rises into the **atmosphere**. As the air moves upward, it cools. The water in the air changes from a gas into liquid drops that form bands of clouds. Earth's rotation causes the clouds to spin. They circle faster and faster around the center of the storm.

Hurricane John

Hurricane Patricia

Hurricane, Typhoon, or Cyclone?

Hurricanes don't only form over the Atlantic and eastern Pacific Oceans. They can form in tropical waters north and south of Earth's equator anywhere on the planet. But hurricanes that form in places other than the Atlantic and eastern Pacific go by different names. In the western Pacific these storms are called typhoons. In the southern Pacific and Indian Oceans they are called cyclones.

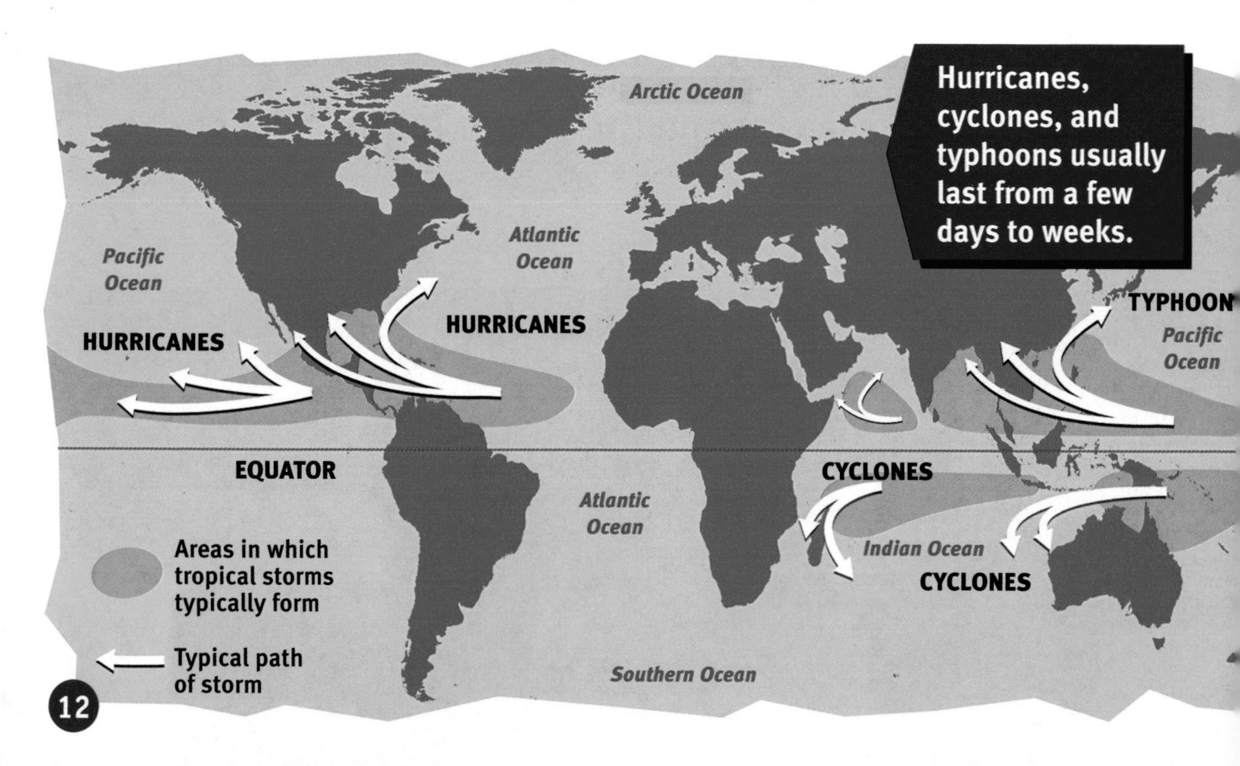

Inside the Storm

The longer hurricanes travel over the ocean, the stronger and bigger they become. Take a look inside one of these colossal storms. Hurricanes are typically 300 miles (483 km) wide.

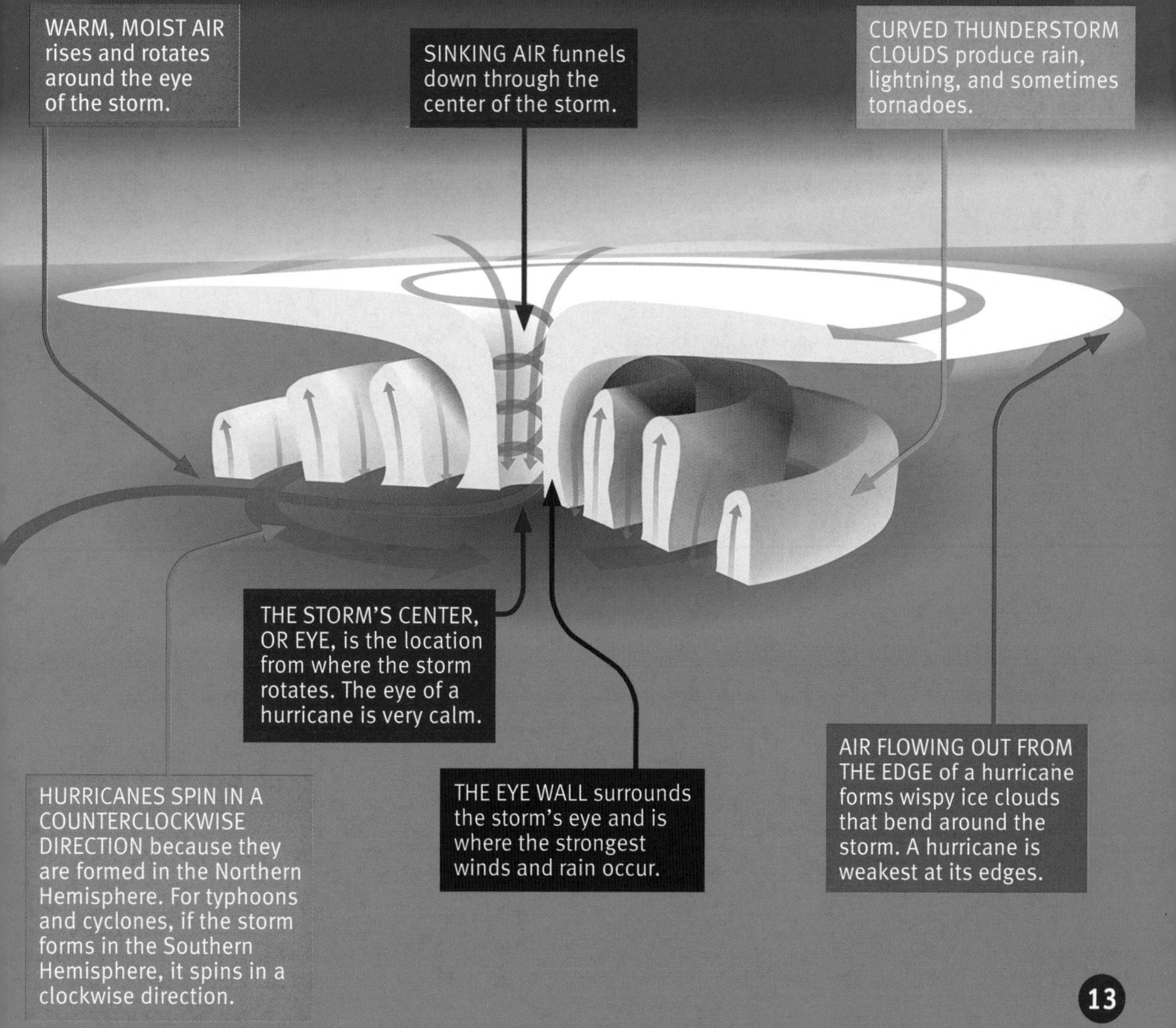

WARM, MOIST AIR rises and rotates around the eye of the storm.

SINKING AIR funnels down through the center of the storm.

CURVED THUNDERSTORM CLOUDS produce rain, lightning, and sometimes tornadoes.

THE STORM'S CENTER, OR EYE, is the location from where the storm rotates. The eye of a hurricane is very calm.

HURRICANES SPIN IN A COUNTERCLOCKWISE DIRECTION because they are formed in the Northern Hemisphere. For typhoons and cyclones, if the storm forms in the Southern Hemisphere, it spins in a clockwise direction.

THE EYE WALL surrounds the storm's eye and is where the strongest winds and rain occur.

AIR FLOWING OUT FROM THE EDGE of a hurricane forms wispy ice clouds that bend around the storm. A hurricane is weakest at its edges.

This specially equipped airplane is flying in the eye, or center, of Hurricane Floyd, which hit the Bahamas and the East Coast of the United States in 1999.

Pilots have been flying missions to monitor conditions inside hurricanes since the 1940s.

CHAPTER

2

Hurricane Trackers

Scientists called **meteorologists** work to predict which tropical storms will become hurricanes. To watch for and monitor the storms, meteorologists use high-tech tools. They then use the information they gather to figure out where hurricanes are headed. This helps keep people in the storm's path safe. Which tools do these scientists use? Let's find out!

Eyes in the Sky

Weather **satellites** are the main way meteorologists track hurricanes. These instruments orbit hundreds to thousands of miles above Earth. Some satellites circle the planet. Others hover in one location to constantly monitor a specific area. The satellites help scientists observe Earth's atmosphere, allowing them to see hurricanes from above. These images can tell scientists about a storm's direction, speed, and size.

Some weather-tracking satellites zoom through space at 7,000 miles per hour (11,265 kph)!

There are dozens of weather satellites orbiting Earth. They monitor all types of severe weather.

Hurricane Hunters fly their planes through a storm four to six times each mission.

Into the Storm

To learn what is happening inside a hurricane, trained pilots fly airplanes carrying scientists into the storms. As their aircraft passes through a hurricane, its instruments take readings on the surrounding weather conditions. The scientists also release sensors attached to small parachutes. The sensors (pictured above) drift through the storm, collecting data on pressure, temperature, wind speed, and **humidity**. The data is sent back to a computer on the airplane.

Classifying Hurricanes

Meteorologists use the Saffir-Simpson Hurricane Wind Scale, shown below, to assign each hurricane a category based on its wind speed.

CATEGORY	WIND SPEEDS	DAMAGE	
1	**74–95 mph** (119–153 kph)	MINIMAL	Buildings mostly unharmed; mobile homes and trees possibly damaged
2	**96–110 mph** (154–177 kph)	MODERATE	Buildings' roofs, windows, and doors damaged; uprooted trees; some flooding
3	**111–129 mph** (178–208 kph)	MAJOR	Buildings damaged; mobile homes destroyed; trees, power lines, and street signs blown down; some flooding
4	**130–156 mph** (209–251 kph)	EXTREME	Roofs and walls collapse; trees, power lines destroyed; some flooding; power outages; people asked to leave coastal areas
5	**more than 157 mph** (more than 252 kph)	CATASTROPHIC	Large buildings damaged; trees, small structures, cars destroyed; lots of flooding; power outages; people asked to leave coastal areas

Sea Sensors

Across the ocean, more than 1,000 floating buoys help scientists keep track of changing weather at sea as hurricanes develop. Buoys in shallow water near the shore are anchored in place. Farther out at sea, drifting buoys move with the ocean's currents. The instruments constantly collect data. They keep track of the height of waves, sudden increases in wind speed, and ocean and air temperatures. This information is sent to weather stations on land.

Ships can call a special phone number to access data collected by buoys and find out what the weather conditions are.

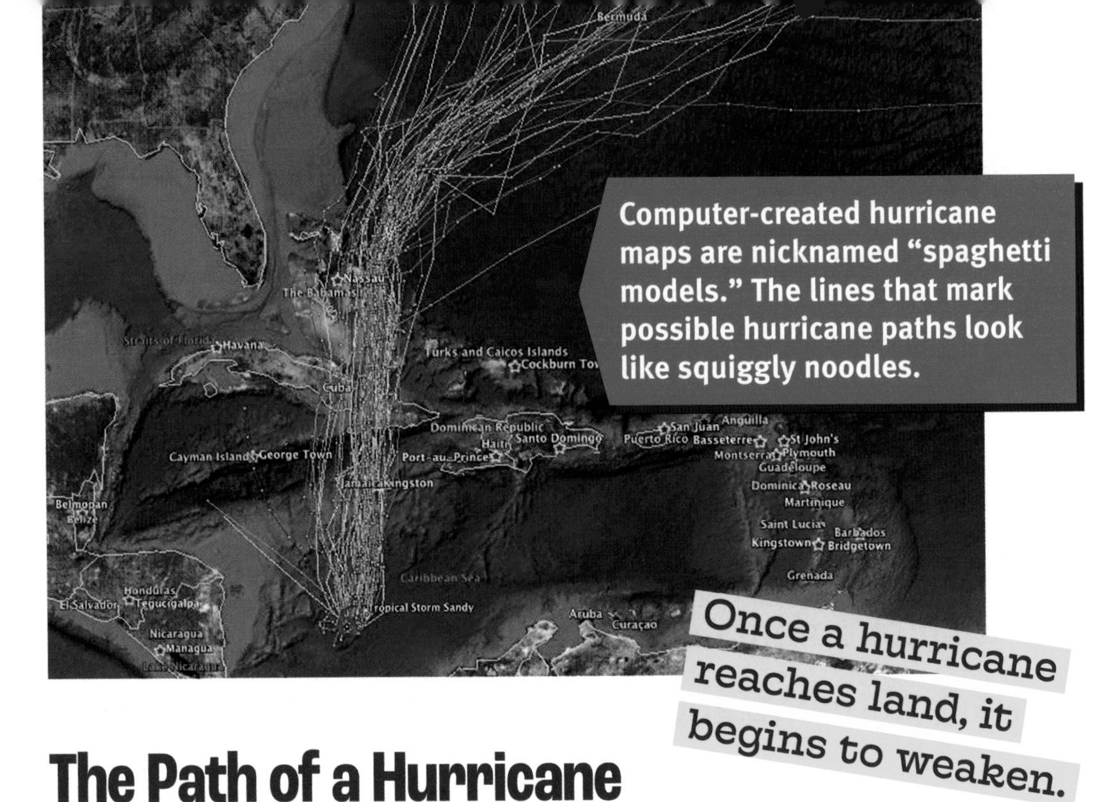

Computer-created hurricane maps are nicknamed "spaghetti models." The lines that mark possible hurricane paths look like squiggly noodles.

Once a hurricane reaches land, it begins to weaken.

The Path of a Hurricane

To predict the path of a hurricane, scientists enter data collected from space, air, and sea into a powerful computer. The computer makes models that predict where a storm is most likely to go. Predicting where a hurricane might make landfall allows scientists to warn people in the storm's possible path. Then, local officials can give **evacuation** orders, if necessary, to move people out of harm's way before a hurricane strikes.

Tools of the Trade

Meteorologists use many tools to track hurricanes, including satellites, airplanes, buoys, and other technologies like the ones below.

1. **Thermometer:** Used to measure air temperature.

2. **Barometer:** Used to measure air pressure.

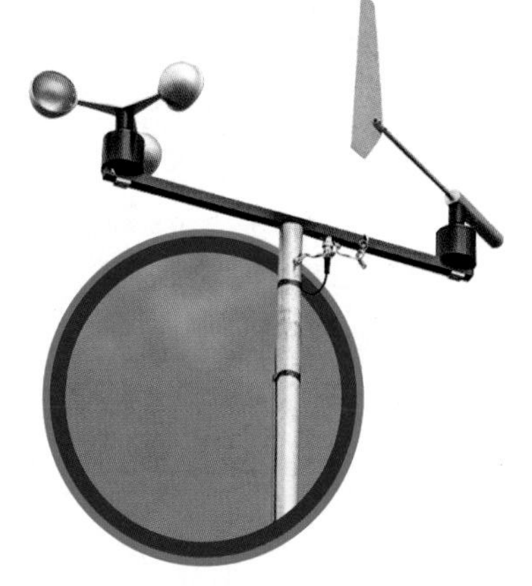

3. **Anemometer:** Spinning device that measures wind speed and direction.

4. **Doppler radar:** Instrument that sends radio waves into the sky to detect rain.

5. **Computer:** Used to analyze data and run models to predict hurricanes.

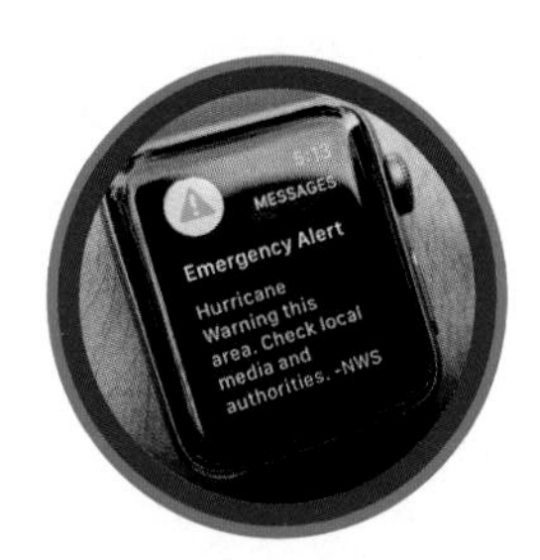

6. **Notification systems:** Social media, email, texts, and apps are used to give warnings.

The BIG Truth

Are Hurricanes Getting More Dangerous?

Yes. Scientists think hurricanes are getting stronger because of **climate change**. People rely on fossil fuels, such as coal, oil, and natural gas, to power vehicles and create electricity. Burning these fuels releases **greenhouse gases**. These gases act like a blanket around the planet, trapping heat. This has caused temperatures on Earth to get hotter, and weather patterns around the world to change. How is climate change exactly affecting hurricanes? Read on!

Hurricanes Cause More Flooding: Climate change is causing ice at Earth's poles to melt. The water pours into the oceans, causing sea levels to rise. Higher seas increase the height of storm surges during hurricanes, so coastal areas could see even more flooding.

Stronger Hurricanes are Becoming More Frequent:

Climate change is also causing oceans to heat up. Warmer waters in the tropics provide more fuel for hurricanes. As a result, scientists have seen a rise in the strongest Category 4 and 5 storms.

Hurricanes Last Longer and Are Wetter: Warmer temperatures created by climate change are allowing hurricanes to hold on to water, or moisture, and moisture is what these storms need to survive. This extra moisture is giving hurricanes energy to last longer, even after they reach land. It is also making these hurricanes produce more rainfall.

In the United States, tropical storms and hurricanes are the only weather events given their own names.

In 2012, Hurricane Sandy washed the Jet Star roller coaster in Seaside Heights, New Jersey, into the Atlantic Ocean.

CHAPTER

3

History-Making Storms

Sometimes hurricanes come and go without doing much damage. They either are not strong enough or they break up quickly over land. But some have caused huge amounts of destruction. High winds from storms have leveled buildings, trees, and power lines. They often leave people without electricity for weeks or months. Their storm surges have swamped entire cities. These are some of the worst hurricanes in recent history.

Hurricane Katrina

On August 29, 2005, Hurricane Katrina hit New Orleans, Louisiana. As a Category 3 hurricane, its winds were not the strongest. But it brought lots of rain and a huge storm surge. Water overflowed or broke through floodwalls, called levees, that surrounded the city. Within a few days, 80 percent of New Orleans was underwater. More than a million people evacuated before the storm, but about 200,000 others were left stranded.

Katrina is the costliest hurricane on record.

New Orleans was underwater after Hurricane Katrina. Many residents lost their homes and had no choice but to move to other parts of the country.

Naming Hurricanes

In the 1950s, meteorologists began naming hurricanes. They thought it would make it easier for people to keep track of the storms during weather forecasts and news reports. The names come from a list created by the World Meteorological Organization. There is one list for Atlantic hurricanes and one for Pacific hurricanes. The names are assigned to tropical storms and hurricanes alphabetically in the order they form. In the Atlantic, no names start with Q, U, X, Y, or Z. In the eastern Pacific Ocean, X, Y, and Z are used.

Storm names can be reused after a period of six years. But if a hurricane is extremely deadly or damaging, its name is retired.

Retired Hurricane Names

HAZEL (1954)
IONE (1955)
DONNA (1960)
BEULAH (1967)
FIFI (1974)
GRETA (1978)
KLAUS (1990)
BOB (1991)
HORTENSE (1996)
FLOYD (1999)
FABIAN (2003)
IVAN (2004)
KATRINA (2005)
GUSTAV (2008)
IGOR (2010)
SANDY (2012)
INGRID (2013)
HARVEY (2016)
IRMA (2017)
MARIA (2017)
FLORENCE (2018)
MICHAEL (2018)

Hurricane Sandy

In October 2012, Hurricane Sandy hit the islands of the Caribbean. Then it moved up the East Coast of the United States, affecting states from Florida to Maine. But the hardest-hit areas were New Jersey and New York City. The Category 1 storm brought storm surges of more than 13 feet (4 m), flooding seaside towns and parts of lower Manhattan. The hurricane destroyed or damaged about 650,000 homes in the United States.

Sandy caused major flooding of New York City's subways, roads, and tunnels.

Harvey dumped one trillion gallons of rain on Houston over four days.

Harvey lasted a record-setting 117 hours after making landfall.

Hurricane Harvey

Hurricane Harvey made landfall in Texas on August 25, 2016, as a Category 4 hurricane. Instead of moving inland, the storm stalled over the coast. For four days, Harvey dumped a record amount of rain on the region. Some areas saw as much as 5 feet (1.5 m), leading to lots of flooding. Rising waters swamped one-third of the city of Houston. As a result of the storm, 68 people died and 300,000 buildings and 500,000 vehicles were damaged.

Hurricanes Irma and Maria

Irma hit Florida as a Category 4 hurricane on September 10, 2017. It ripped roofs off homes and flooded coastal cities. The storm knocked out power and forced six million people to evacuate. Before striking the United States, Irma caused damage to several islands in the Caribbean. Puerto Rico was still reeling from Irma when Maria, another Category 4 hurricane, hit two weeks later. Maria caused widespread destruction and killed almost 3,000 people in Puerto Rico.

Most people in Puerto Rico did not have access to electricity for several months after Maria struck.

Haiyan destroyed between 70 and 80 percent of all the structures in its path.

About 16 million people were affected by damage caused by Typhoon Haiyan.

Typhoon Haiyan

On November 8, 2013, Typhoon Haiyan made landfall in the Philippines, a country in Southeast Asia. At landfall, the storm's wind speeds were among the strongest ever recorded, reaching 195 miles per hour (314 kph). That puts it in the same class as a Category 5 hurricane. Its strong winds flattened entire towns. It also brought a massive storm surge that washed away other towns. Many people lost their lives.

Hurricanes can spawn severe, short-lived windstorms called tornadoes.

Concrete, dome-shaped homes in Florida are built to stand up to a hurricane's strong winds.

CHAPTER

4

Weathering Hurricanes

Many coastal towns are working to bring back natural areas and native plants. Why? They offer natural flood protection against hurricanes and slow down storm surges before they reach communities. Other places have built artificial barriers to hold back floodwaters. Many cities require that houses are built to resist strong winds or high water. This has led to wild new hurricane-proof home designs.

Along many coasts, projects are underway to plant grasses to anchor sand in place and rebuild dunes.

Natural Barriers

Reefs, sand dunes, and mangrove forests are natural barriers against storm surges. So are **wetlands**. These swampy areas near waterways soak up extra water from storms. But human development has destroyed many of these natural defenses. Buildings and pavements do not allow water to drain. Today, many coastal areas are bringing back wetlands and sand dunes to lessen the impacts of hurricanes.

Artificial Barriers

People have come up with many ways to hold back water during hurricanes. They have built levees along waterways and seawalls along ocean fronts. They have installed pumps to remove large amounts of stormwater. Canals and ponds are also built to drain and contain water during storms. If a hurricane is powerful enough, though, these structures can fail. This has left engineers looking for better ways to keep floodwaters away.

Water floods over a levee in New Orleans, Louisiana, after Hurricane Katrina.

Weather Alert!

In the United States, the National Hurricane Center issues weather alerts before tropical storms and hurricanes. Here's how to prepare and respond if there is a watch or warning for your area.

WEATHER ALERT	
Tropical Storm or Hurricane WATCH	Tropical Storm or Hurricane WARNING
MEANING	
Tropical storm or hurricane conditions could reach an area within 48 hours.	Tropical storm or hurricane conditions are expected to reach an area within 36 hours.
WHAT YOU SHOULD DO	
Stay up-to-date on changing storm conditions.	Listen to local news reports for storm updates and check alerts on your smart devices.
Stock up on emergency supplies: canned food, bottled water, and a first aid kit.	Double-check supplies. Don't wait until the last minute to stock up on necessities.
Prepare an emergency plan and know how to contact other people and where to go if you need to leave the area.	Keep cell phones charged in case the storm causes a power outage. Make sure your car is filled with gas.
Park cars in the garage and bring in outdoor items that could blow away, such as patio furniture.	During the storm, stay away from windows on the first floor of your home. If there is flooding, move away from the first floor.
	If told to evacuate, leave immediately.

Workers began building the Maeslantkering in 1991 and completed it in 1997.

New Orleans has 350 miles (563 km) of floodgates and levees to protect against storm surges.

Floodgates

In 1953, a hurricane-like storm flooded part of the Netherlands in Europe. To help protect the area, the country decided to build the Maeslantkering (MAHS-lawnt-keh-ring). This giant gate protects the city of Rotterdam from storm surges. It stays open so ships can pass, but it can be closed during storms. This gate was the inspiration for similar structures in hurricane-prone areas, including New Orleans, Louisiana.

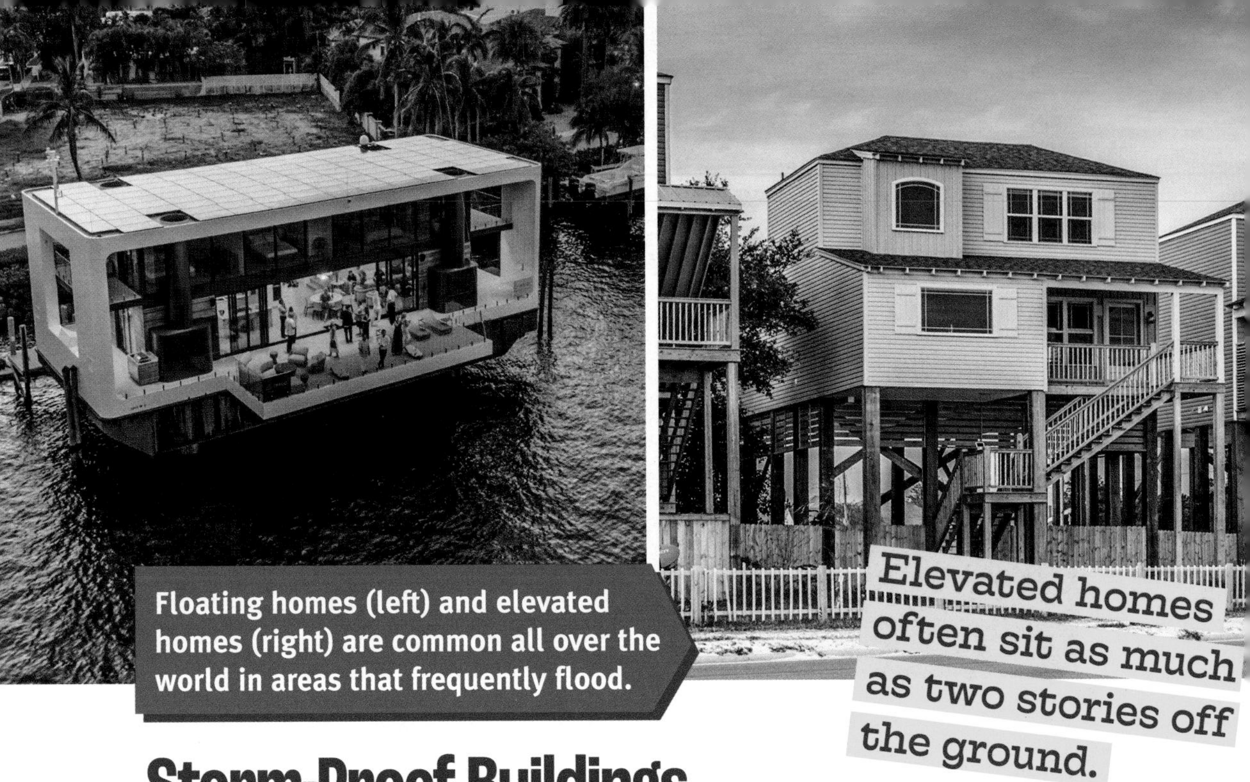

Floating homes (left) and elevated homes (right) are common all over the world in areas that frequently flood.

Elevated homes often sit as much as two stories off the ground.

Storm-Proof Buildings

In many areas that get hurricanes, homes can be built to stand up to the storms. Windows have stronger glass, so they are less likely to break. Storm shutters on windows can be closed to block strong winds. Homes can also be elevated to protect against rising waters. Engineers have designed floating homes, too. They can survive flooding. Domed houses have also been built to better withstand strong winds.

A Stormy Future?

Because of climate change, stronger hurricanes with more rain could become the norm. But many people are already making a difference. How? By using **clean energy** such as solar power. Researchers are also thinking of new and better ways to reduce hurricane damage. You can help, too, by reducing greenhouse gases in your daily life. For example, ride your bike or walk whenever you can. All of our small steps together will make a big difference!

A sign over a highway gives people early warning that a hurricane is coming.

Hurricane Strikes

Scientists began keeping records of hurricanes in 1851. The data has revealed where hurricanes strike most often. Study the bar graph showing the top five states in the United States where the most hurricanes have made landfall, and then answer the questions that follow.

Analyze It!

1. Based on the data in the graph from 1851 to 2020, is the East or West Coast most at risk from hurricanes?
2. Florida experiences about twice as many hurricanes as Texas—the second-hardest-hit state. Why do you think that is?
3. What other information would you need to determine which state suffered the biggest impact from hurricanes during this period?

Top Five States Hit by Hurricanes

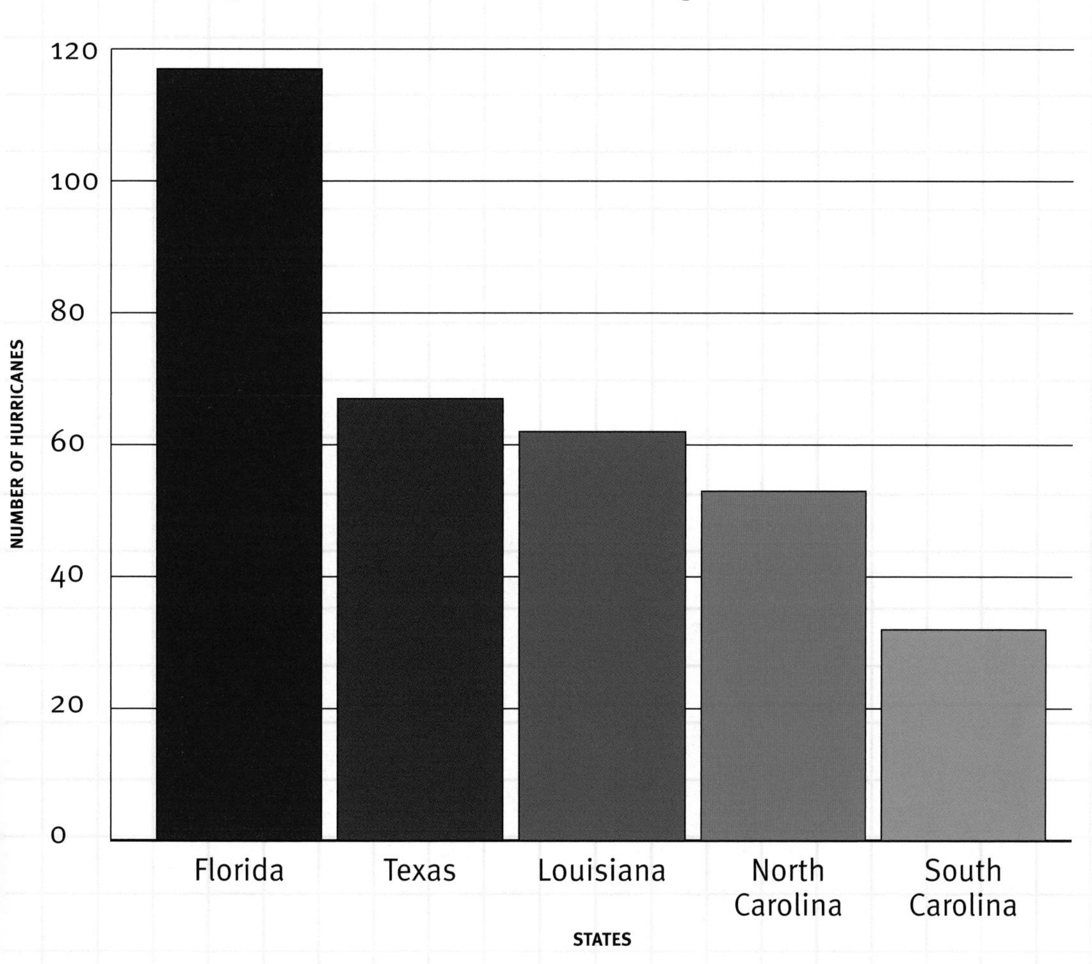

Source: National Hurricane Center
Data is for years 1851–2020.

ANSWERS: 1. East Coast 2. It sticks out into the Gulf of Mexico and Atlantic Ocean, putting it in the direct path of many hurricanes. 3. The strength of the hurricane that hit the states, cost of damages, and death toll.

Design a Hurricane-Proof House

Activity

Engineers are working to make homes stronger in order to withstand severe storms. Design, build, and test your own hurricane-proof house.

Materials

Building materials: cardboard, clay, construction paper, craft sticks, glue, scissors, straws, tape, toothpicks

Pencil

Paper

Tray or pan

Water

Straw

Directions

1. Gather building materials from the list above that you can find around your home. Ask an adult before using them.

2. How can you use the materials to build a hurricane-proof house? It needs to stand up to strong winds and rising water. Draw your design on paper.

3 Use the building materials to build a model of your hurricane-proof house inside the tray.

4 Fill the tray halfway with water.

5 Use the straw to blow as hard as you can on your house and the water. What happens to your model? How can you improve your design?

Explain It!

Using what you learned in the book, can you explain what happened and why? If you need help, turn back to page 38.

True Statistics

The year the National Hurricane Center officially started naming storms: 1953

Most tropical storms or hurricanes to form in the Atlantic and Pacific Oceans at the same time: 6 (1992 and 2019)

Length of hurricane season: 6 months (June to November)

Percentage of hurricanes that strike the United States and impact Florida: 40 percent

Number of Category 5 hurricanes to hit the United States in the past 100 years: 5

The width of a typical hurricane: 300 miles (483 km)

Most tornadoes spawned from a hurricane: 120 confirmed tornadoes spawned by Hurricane Ivan in 2004

Did you find the truth?

F The wind speeds of a Category 4 hurricane can be up to 145 miles per hour (233 kph).

T Scientists fly into hurricanes to collect data.

Resources

Other books in this series:

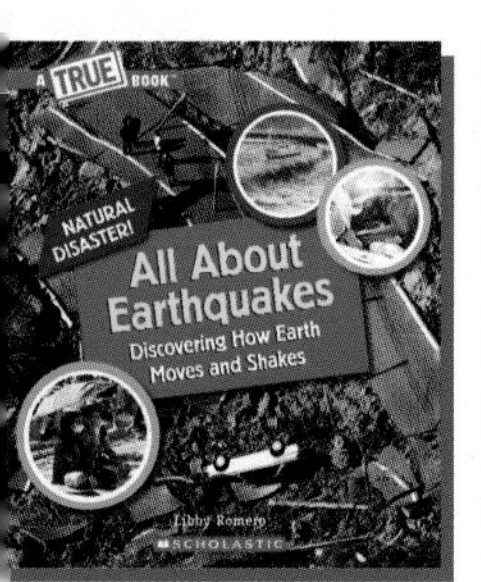

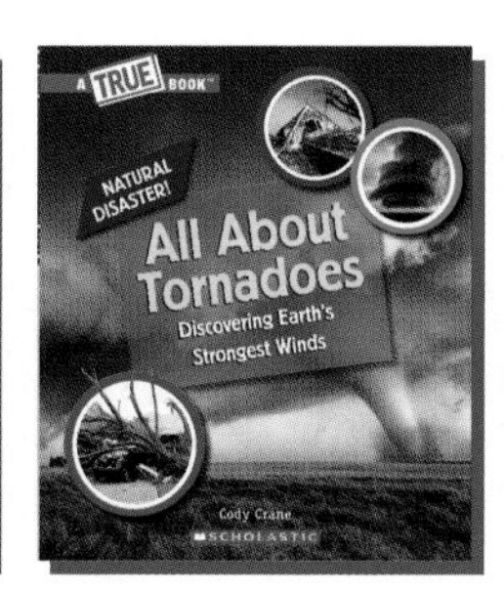

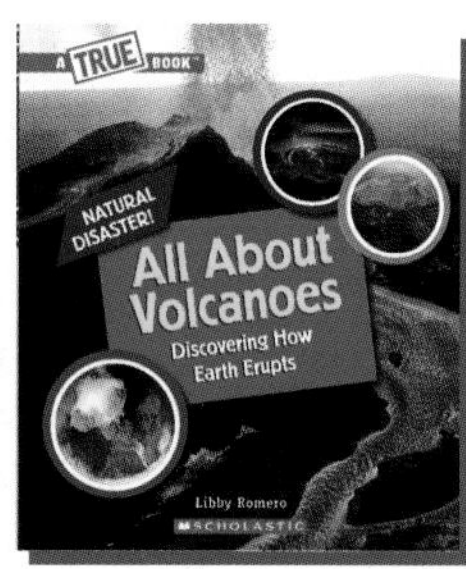

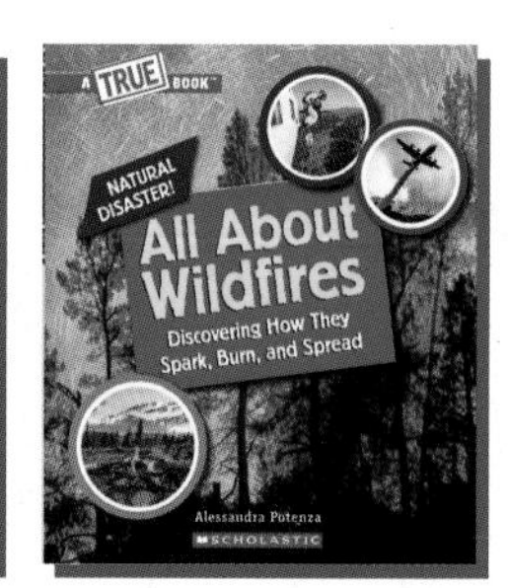

You can also look at:

Carson, Mary Kay. *Inside Hurricanes.* New York: Sterling, 2010.

Gregory, Josh. *The Superstorm Hurricane Sandy.* New York: Children's Press, 2013.

Rogers, Lisa Waller. *The Great Storm: The Hurricane Diary of J. T. King, Galveston, Texas, 1900*. Lubbock, Texas: Texas Tech University Press, 2010.

Simon, Seymour. *Hurricanes.* New York: HarperCollins, 2007.

Tarshis, Lauren. *I Survived Hurricane Katrina, 2005.* New York: Scholastic, 2011.

Tarshis, Lauren. *I Survived the Galveston Hurricane, 1900.* New York: Scholastic, 2021.

Glossary

atmosphere (AT-muhs-feer) the mixture of gases that surrounds a planet

clean energy (kleen EN-ur-jee) energy from natural sources, such as the sun or wind, that do not run out or produce pollution

climate change (KLYE-mit chaynj) an increase in temperatures on Earth and shifts in global weather patterns

evacuation (ih-VAK-yoo-A-shun) moving away from an area or building because it is dangerous there

evaporates (ih-VAP-or-atez) changes from a liquid to a gas

greenhouse gases (GREEN-howss gas-ez) gases that trap heat in a planet's atmosphere, causing average global temperatures to rise

humidity (hew-MID-i-tee) moisture in the air

hurricane (HUR-ih-cane) a rapidly spinning tropical storm

meteorologists (mee-tee-uh-RAH-luh-jists) scientists who study the weather

satellites (SAT-uh-lights) natural or human-made objects that orbit a planet

storm surge (storm surj) water pushed toward shore by a hurricane

wetlands (WEHT-landz) lands covered with shallow water

Index

Page numbers in **bold** indicate illustrations.

About the Author

Cody Crane is an award-winning children's writer, specializing in nonfiction. She studied science and environmental reporting at New York University. She always wanted to be a scientist but discovered that writing about science could be just as fun as doing the real thing. She lives in Houston, Texas, with her husband and son.

Photos ©: cover top left: Nathan Blaney/Getty Images; cover top right: Warren Faidley/Getty Images; cover bottom: Juan DeLeon/Icon Sportswire/Getty Images; back cover: NOAA GOES Project/Getty Images; 3: NOAA GOES Project/Getty Images; 4: Warren Faidley/Getty Images; 5 top: Joe Raedle/Getty Images; 6–7: Pictures From History/age fotostock; 8–9: Warren Faidley/Getty Images; 10 left: Photo Josse/Bridgeman Images; 10 right: Robert W Kelley/The LIFE Picture Collection/Getty Images; 11 left: NOAA/Wikimedia; 11 right: Brett Gundlock/Getty Images; 12: NASA; 13: Kelvinsong/Wikimedia; 14–15: Chris Sattlberger/Anzenberger/Redux; 17 left: AC NewsPhoto/Alamy Images; 17 right: NCAR/UCAR/NASA; 19: Michael Dwyer/Alamy Images; 20: Google Earth; 21 top left: Stephen J. Krasemann/Science Source; 21 top right: vuk8691/Getty Images; 21 bottom left: Moreno Soppelsa/Dreamstime; 21 bottom center: Mario Villafuerte/Bloomberg/Getty Images; 21 bottom right: Richard Unten/Flickr; 22–23 background: Chip Somodevilla/Getty Images; 22 bottom: Paul Souders/Getty Images; 23 top: NOAA GOES Project/Getty Images; 23 bottom: Erich Schlegel/Getty Images; 26: FEMA/Alamy Images; 27: MODIS Rapid Response Team/GSFC/NASA; 28: Michael Bocchieri/Getty Images; 29: Joe Raedle/Getty Images; 30: Mario Tama/Getty Images; 31: Images & Stories/Alamy Images; 32–33: Andrew J. Whitaker/The Post And Courier/AP Images; 34: Bill Brooks/Alamy Images; 35: POOL/AFP/Getty Images; 37: frans lemmens/Alamy Images; 38 right: BHammond/Alamy Images; 39: Robert D. Barnes/Getty Images; 40–41, 42 graph paper: billnoll/Getty Images; 42–43 illustrations: Gary LaCoste; 44: Nathan Blaney/Getty Images.

All other photos © Shutterstock.